CIRIA C626   London 2004

CD ENCLOSED

Model agreements for sustainable water management systems

Model agreement for rainwater and greywater use systems

P Shaffer   CIRIA

C Elliott   Environment Agency

J Reed   Atkins

J Holmes   Weightman Vizards

M Ward   Atkins

CIRIA *sharing knowledge ▪ building best practice*

Classic House, 174–180 Old Street, London EC1V 9BP, UK
TEL +44 (0)20 7549 3300   FAX +44 (0)20 7253 0523
EMAIL enquiries@ciria.org
WEBSITE www.ciria.org

Summary

This guide provides basic advice on the use and development of model operation and maintenance agreements for rainwater and greywater use systems alongside simple guidance on their incorporation in developments. It identifies maintenance considerations and provides an outline of how the long-term responsibilities for the maintenance of rainwater and greywater use systems can be allocated.

Model agreements for sustainable water management systems. Model agreement for rainwater and greywater use systems

CIRIA

Shaffer, P; Elliott, C; Reed, J; Holmes, J; Ward, M

CIRIA C626   © CIRIA 2004   ISBN 0-86017-626-6   RP664

Environment Agency Technical Report P2-261/19/TR

British Library Cataloguing in Publication Data

A catalogue record is available for this book from the British Library.

Keywords		
Environmental good practice, sustainable construction, water resources, water quality		
Reader interest	**Classification**	
Developers, building-owners, consulting engineers, water utilities, local authorities, architects, environmental regulators, planners, contractors, property-owners, water conservation enthusiasts, housing associations, facilities managers	AVAILABILITY	Unrestricted
	CONTENT	Technical guidance and model agreement
	STATUS	Committee-guided
	USER	Developers, architects, engineers, regulators

Published by CIRIA, Classic House, 174–180 Old Street, London EC1V 9BP, UK.

Background to the guide

This guide has been prepared for use by all organisations involved in the provision and maintenance of sustainable water management systems. This may include:

- owners and developers
- landscape architects
- consulting engineers
- land use planners
- architects
- environmental regulators
- sewerage undertakers
- residents
- tenants
- facility managers
- property-owners and landowners.

Model agreements for sustainable water management systems is a series of documents resulting from CIRIA Research Project 664, which was undertaken to facilitate and encourage the long-term maintenance of sustainable drainage systems (SUDS) and rainwater/greywater use systems through the development and application of model agreements. A model agreement is just one method of allocating responsibilities for the maintenance of systems and consists of a legal agreement that can be used as the basis for agreements between two or more parties.

The series comprises two main outputs.

Model agreement for rainwater/greywater use systems **(CIRIA C626).** This guide explains the background to the operation and maintenance of rainwater/greywater use systems, providing a framework for the long-term operation and maintenance of reuse systems. The model agreement covers a variety of scenarios, from single properties to multi-occupancy properties.

Model agreements for SUDS **(CIRIA C625).** This sister publication provides background information on, and a long-term framework for, the operation and maintenance of sustainable drainage systems (SUDS). Model agreements were developed for specific scenarios:

- implementation and maintenance of SUDS, either as a planning obligation under Section 106 of the Town and Country Planning Act 1990 or as a condition attached to planning permission
- implementation and maintenance of SUDS between two or more parties (outside the requirements for planning permission), ie a private SUDS model agreement.

These two publications are not intended for use as extensive reference works on the design and construction of sustainable water management systems but complement existing guidance documents and frameworks, full details of which are given in Chapter 1.

Acknowledgements

Research contractor This guidance document is the outcome of Research Project 664. It was prepared by **CIRIA, Atkins** and **Weightman Vizards**.

Authors

Paul Shaffer BSc
Paul is a project manager with CIRIA and for eight years has encouraged and implemented the sustainable use and management of water in the built environment. He is responsible for a number of projects designed to help overcome the perceived technical and management barriers to the sustainable management of water.

Craig Elliott BA MSc
Craig is a policy manager for the Environment Agency with responsibility for flood risk management research and development. Before this he was an associate at CIRIA and has more than 16 years' experience of research into water sustainability issues.

Jon Reed BSc MSc CEng MICE MCIWEM
Jonathan is a chartered civil and environmental engineer whose experience spans water resources, sustainable development and infrastructure design. He is responsible for feasibility and design work for SUDS solutions for both new-build and retrofit situations. He has also represented developers and local authorities with respect to the implementation of SUDS through the planning process.

John Holmes BA LLB
John is a partner at Weightman Vizards Solicitors where he is head of Commercial Property, Planning, Environmental and Construction. Based in Manchester and an acknowledged specialist in his field, he has long experience of the public sector. This expertise extends to advising on all aspects of planning and environmental law.

Martin Ward BEng MSc MPhil CEng MIMechE
Martin is a chartered mechanical engineer with more than 10 years' experience of the water industry. He spent eight years in Anglian Water's Research and Development Department, where he researched new water treatment and recycling processes and was involved in the practical application of rainwater, greywater and blackwater recycling projects. Currently, he is a project manager with Atkins Ltd, working mainly on the renovation and design of sewage treatment works.

Steering group Following CIRIA's usual practice, the research project was guided by a steering group, as listed below.

Chair	Charles Ainger	MWH
Attending members	David Barraclough	Royal Town Planning Institute
	Nick Cooper	Alderburgh Ltd
	Rupert Cowen	Hammond Suddards Edge
	Mark Everard	The Natural Step/Environment Agency
	Ian Hardwick	JJ Gallagher
	Paul Jeffrey	Cranfield University
	Denis Lane	Stevenage Borough Council
	Rebecca Lemon/	
	David Calderbank	Environment Agency
	Lawrence Mbugua	Davis Langdon Consultancy (DTI representative)
	Alex Middleton	The Greenbelt Group of Companies Ltd
	Terry Nash	Gusto Construction Ltd

John Nicholson	Severn Trent Water	
Stan Redfearn	The BOC Foundation	
Andrew Shuttleworth	SEL Environmental	
Neil Smith	NHBC	
Chris Tyler	WSP Development Ltd	
Tom Wild	SEPA	
Steve Wilson	Environmental Protection Group Ltd	

Corresponding members

Nick Beckwith/Marc Haley	Wardell Armstrong
Philip Day	Severn Trent Water Ltd
Steve Dickie	Entec Ltd
Richard Kellagher	HR Wallingford Ltd
Robin Mynard	Defra
Simon Walster	Ofwat

CIRIA managers CIRIA's research managers for the project were **Paul Shaffer** and **Craig Elliott**.

Project funders The project was funded by:

CIRIA Core Programme
Department of Trade and Industry
The BOC Foundation
Environment Agency
Severn Trent Water
Ofwat
NHBC
WSP Development
SEL Environmental
Environmental Protection Group Ltd

Contributors CIRIA and the research contractors wish to acknowledge the following individuals who provided information, help and feedback for the project and whose contributions have been important to the development of the model agreements:

Ken Banfield	Anglian Water
Bob Bray	Robert Bray Associates
Dave Brook	ODPM – Planning
Reg Brown	BSRIA
David Butler	Imperial College
Jim Conlin	Scottish Water
Denis Cooper	Ipswich Borough Council
Robert Cunningham	The Wildlfe Trusts
Graham Fairhurst	Borough of Telford & Wrekin
David Gallagher	Environment Agency (greywater trialist)
Carolin Gohler	Cambridge City Council
Nick Grant	Elemental Solutions
Gill Greatorex	Environment Agency (greywater trialist)
John Hamilton	Northern Ireland Water Service
David Harley	SEPA
Cath Hassell	Construction Resources
Andy Hawkes	JJ Gallagher
Sian Hills	Thames Water
Mark Holland	SANDS
Brian Hurst	Freewater UK
Kendrick Jackson	formerly Gleeson Homes, now House Builders Federation
Chris Jefferies	University of Abertay

Phil Jobson	Welcome Break
Peter Johns	Formpave
Mike Johnson	ODPM
Bruce Kavanagh	Tully D'Ath
Chris Kearns	South Gloucestershire Council
David Knaggs	Metropolitan Water Company
Denis Lane	Stevenage Borough Council
Richard Lemon	Hampshire CC
Kevin Light	Davies Light Associates
Kirsteen Macdonald	Ewan Associates
Chris Mackenzie	Harborough District Council
Prosper Paul	Environment Agency
Stella Peterson	Salford City Council
Neil Robinson-Welsh	Drainage Management
Owen Saward	Wealden District Council
David Sellers	Leeds City Council
Chris Shirley-Smith	Water Works UK
Nick Trollope	Fairview New Homes
Mike Waite	Defra
Adrian Watkins	CEIMA Ltd
Joe Whiteman	Countryside Properties
Paul Williams	Gramm Environmental
Peter Woods	Tesco.

Other contributors CIRIA and the research contractors would also like to acknowledge the following organisations for their help in preparing the model agreements:

National SUDS Working Group
UK Rainwater Harvesting Association.

Contents

Figures and tables

FIGURES

TABLES

Glossary

AA air gap – A device that uses an air gap to protect the mains potable water supply from contamination by a Class 5 risk (see Water Supply (Water Fittings) Regulations 1999).

Blackwater – Effluent discharged as sewage containing faecal matter.

Coliforms – A group of bacteria found in the intestines, faeces (of most animals), nutrient-rich waters, soil and decaying plant material.

Commuted sum – A single payment paid at the beginning of an agreement to cover maintenance for an agreed period of time.

Disinfection – Treatment of water to remove reduce infective risk

Environmental regulators – The Environment Agency in England and Wales, SEPA in Scotland, and the Northern Ireland Environment and Heritage Service in Northern Ireland.

Filtration – The act of removing sediment or other particles from a fluid by passing it through a filter.

First flush – The initial runoff from a site or catchment following the start of a rainfall event. As runoff travels over a catchment it will collect or dissolve pollutants and the "first flush" portion of the flow may be the most contaminated as a result. This is especially the case for intense storms and in small or more uniform catchments. In larger or more complex catchments pollution wash-off may contaminate runoff throughout a rainfall event and the first flush may not occur.

Greywater – Wastewater from sinks, baths, showers and domestic appliances. Kitchen sink or dishwasher wastewater is not generally collected for use because it has high levels of contamination from detergents, fats and food waste, making filtering and treatment difficult and costly.

Greywater use system – An above- or below-ground system that collects, stores, treats and disinfects greywater for use as reclaimed water in properties.

Impermeable surface – An artificial non-porous surface that generates a surface water runoff after rainfall.

Infiltration (to the ground) – The passage of surface water through the surface of the ground.

Model agreement – A legal document that can be completed to form the basis of an agreement between one or more parties regarding the maintenance and operation of sustainable water management systems.

Permeable surface – A surface formed of material that is itself impervious to water but, by virtue of voids formed through the surface, allows infiltration of water to the sub-base through the pattern of voids – for example, concrete block paving.

Pervious surface – A surface that allows inflow of rainwater into the underlying construction or soil.

Rainwater harvesting system – An above- or below-ground storage system that collects, treats, stores and distributes runoff of rain or snow from roofs.

Rainwater use system – A system that collects rainwater from where it falls rather than allowing it to drain away, treats and stores it, then distributes it for use. This includes water collected within the boundaries of a property, from roofs and surrounding surfaces, including areas of hardstanding and pervious paving.

Reclaimed water – Water that has been treated so that its quality is suitable for specified purposes such as irrigation and toilet flushing.

Runoff – Water flow over the ground surface to the drainage system. This occurs if the ground is impermeable, is saturated or if rainfall is particularly intense.

Section 106 TCPA 1990 – A section within the Town and Country Planning Act 1990 that allows a planning obligation to a local planning authority to be legally binding.

Separate sewer – A sewer for surface water or foul sewage, but not a combination of both.

Sewerage undertaker – A collective term relating to the statutory undertaking of water companies that are responsible for sewerage and sewage disposal, including surface water from roofs and yards draining through public sewers.

Source control – The control of runoff or pollution at or near its source.

Sub-catchment – A division of a catchment, allowing runoff management as near to the source as is practicable.

SUDS – Sustainable drainage system: a sequence of management practices and control structures designed to drain surface water in a more sustainable fashion than some conventional techniques.

Sustainable water management system – The collective term for a system that promotes the sustainable management of water. (In this publication, SUDS and rainwater and greywater use systems are the main sustainable water management systems considered.)

Treatment – Improving the quality of water by physical, chemical and/or biological means.

Abbreviations

BOD	biochemical oxygen demand
BSRIA	Building Services Research Information Association
BTSW	Buildings That Save Water
CDM	Construction (Design and Management) Regulations
Cfu	colony-forming units
COSHH	Control of Substances Hazardous to Health
Defra	Department of Environment, Food and Rural Affairs
DTI	Department of Trade and Industry
EA	Environment Agency
HMSO	Her Majesty's Stationery Office
HSE	Health and Safety Executive
NAW	National Assembly for Wales
nm	nanometres
NSWG	National SUDS Working Group
NTU	nephelometric turbidity units
ODPM	Office of the Deputy Prime Minster
Ofwat	Office of Water Services
PPG	Planning Policy Guidance
SUDS	sustainable drainage system
SWMS	sustainable water management system
TCPA	Town and Country Planning Act 1990
USEPA	United States Environmental Protection Agency
UV	ultraviolet (light)
WC	water closet (toilet)
WLC	whole-life costing
WRAS	Water Regulations Advisory Scheme

1 Introduction

Sustainable drainage and rainwater/greywater use systems in buildings form a key part of sustainable developments by reducing the impacts that might otherwise occur to surface water runoff and water resources. Effective systems can potentially cut water consumption, easing pressure on water resources. They can also reduce demand on wastewater systems. In a similar way to SUDS, rainwater systems can help reduce downstream flooding.

CIRIA's recent publications on SUDS and rainwater/greywater use systems have identified the question of eventual ownership of the systems – in particular, who will maintain them – as a major challenge to achieving wider uptake of sustainable water management systems. It is essential to maintain and repair these types of systems properly if they are to perform consistently at design levels, as well as to minimise health and safety hazards.

This publication provides an example model agreement and simple guidance on its implementation within developments. A model agreement is a legal document that can be used as the basis for agreements between two parties (normally the customer and the maintenance provider) for the maintenance of systems.

The model agreement included here is an example of what can be used, although it is not always necessary to use it. The model agreement has also been supplied as an electronic document, so that users may amend the wording and the clauses to reflect specific circumstances.

SCOPE

This guide aims to promote and encourage the sustainable use and management of water within the built environment by providing basic advice on the use and development of maintenance agreements for rainwater and greywater use systems alongside simple guidance on their incorporation into developments. The model agreement developed is relevant to current legislation and policies within England and Wales (at mid-2004). A complementary publication, CIRIA C625, provides similar guidance on model agreements for SUDS.

The specific objectives of providing the model agreements and guidance are to:

- encourage the incorporation of sustainable water management systems in new and existing developments
- help developers and/or practitioners incorporate sustainable water management systems into developments
- establish standard approaches to the allocation of responsibilities for the maintenance of sustainable water management systems
- make the adoption and allocation of maintenance for systems more straightforward so that clients of the construction industry can benefit from cost savings and reduced likelihood of operation and maintenance problems in the future.

SOURCES OF INFORMATION

This guide and the associated model agreement have been developed from an extensive review of legislation and policy in England and Wales and through consultation with stakeholders from the construction and water industries. The model agreement and guidance documents have been reviewed and agreed by a dedicated project steering group of experienced individuals representing a wide range of stakeholders in the sustainable management of water.

STRUCTURE OF THE BOOK

Chapter 1 – Introduction explains the scope of the guidance and the project. It lists the other guidance that can be used to complement this guide and to help in the implementation of the model agreement.

Chapter 2 – Sustainable water management shows how sustainable development can be applied to the water environment and how sustainable water management systems can contribute to sustainable development.

Chapter 3 – Background to rainwater and greywater use gives the context to water resource management and explains the principles of sustainable water use within the built environment.

Chapter 4 – Rainwater and greywater use systems provides background information on the key components of rainwater and greywater use systems and the potential that exists to link rainwater harvesting systems to SUDS.

Chapter 5 – Policy, regulation and guidance sets out details of the regulatory framework for rainwater and greywater use systems. The chapter also outlines some of the guidance available on systems.

Chapter 6 – Maintenance covers the planning and implementation of maintenance for systems.

Chapter 7 – Rainwater and greywater use systems model agreement describes the framework for the model agreement.

Chapter 8 – Commentary on the model agreement explains how the model agreement and schedule may be completed and used.

The model agreement is provided both as a printed booklet and as an MS Word document on CD-ROM, so that readers may adapt it for their own use. The booklet and the CD-ROM are included in the box along with this book. Additionally, an electronic template for the model agreement can be downloaded from <www.ciria.org/suds>.

RELATIONSHIP TO OTHER GUIDANCE

This guide forms part of a suite of CIRIA publications relating to both SUDS and rainwater/greywater use systems. Together they offer detailed information on the design and operation of sustainable water management systems.

Related guidance from CIRIA

Rainwater and greywater use in buildings. Best practice guidance, publication C539 (Leggett *et al*, 2001a). Provides best practice guidance on the use of rainwater and greywater systems in buildings.

Rainwater and greywater use in buildings – decision-making for water conservation, Project Report 80 (Leggett *et al*, 2001b). Presents an overview of the use of rainwater and greywater as a water conservation measure.

Sustainable water management in land use planning, publication C630 (Samuels *et al*, 2004). Sets out guidance on the incorporation of water resource and wastewater treatment issues as part of the planning process for new developments.

Guidance from BSRIA

Rainwater and greywater in buildings – project report and case studies, Technical Note TN 7/2001 (Brewer *et al*, 2001). Gives details of demonstration sites with rainwater or greywater use systems in the UK.

Water reclamation guidance – design and construction of systems using grey water, Technical Note TN 6/2002 (Brown and Palmer, 2002a). Highlights issues of concern affecting the design, construction and installation of greywater use systems.

Water reclamation guidance – laboratory testing of systems using grey water, Technical Note TN 7/2002 (Brown and Palmer, 2002b). Sets out a methodology for establishing the performance of greywater use systems.

Guidance from the Environment Agency

A study of domestic greywater recycling (EA, 1999). Presents information on a trial of greywater use systems.

Conserving water in buildings (water efficiency fact cards) (EA, 2001). Furnishes an overview of measures to conserve water in buildings.

Harvesting rainwater for domestic uses: an information guide (EA, 2003). Provides information on rainwater use systems.

2   Sustainable water management

INTRODUCTION

The concept of sustainable water management supports economic and social development by optimising the use and management of water for people, agriculture, commerce and industry, while protecting and improving the environment for the future.

The Government wants sustainable development to be at the heart of policy-making. The national strategy is defined in *A better quality of life, a strategy for sustainable development in the UK* (DETR, 1999). In answering the question "What is sustainable development?", the strategy states:

> *At its heart is the simple idea of ensuring a better quality of life for everyone, now and for generations to come.*

The strategy specifically identifies water as an example of a renewable resource, which "should be used in ways that do not endanger the resource or cause serious damage or pollution".

Reconciling the water needs of the natural environment with the demands of society poses difficult challenges. The UK environment is under pressure from many directions – increased housing, increased population density (particularly in the South East) and extended road networks – to meet the growing expectations of a population of rising affluence for an improved quality of life. In addition, the realisation and uncertainties of future climate change provide further impetus for the adoption of a precautionary approach to water management.

DEFINING SUSTAINABLE WATER MANAGEMENT SYSTEMS

Sustainable water management systems are those systems or practices that support the sustainable management of water and positively contribute to the goals of sustainable development. This series of guidance is primarily concerned with sustainable drainage systems and rainwater and greywater use systems. In some circumstances, SUDS and rainwater use systems can be combined, but within the UK this practice is not widespread.

DRIVERS FOR SUSTAINABLE WATER MANAGEMENT

Sustainable water management is a concept that includes long-term environmental and social factors in decision-making about the way water is managed or used in the built environment. It considers the quantity and quality of water used and disposed of as well as safeguarding the local environment and amenities. The drivers for sustainable water management are listed in Table 2.1.

Table 2.1   *Drivers for sustainable water management*

Climate change	There is growing evidence that our climate is changing. Household water use may rise as a result of hotter, drier summers, while flood risk could increase because of wetter, colder winters. Climate change may also affect groundwater and river flow regimes. Sustainable water management systems may help alleviate the impacts of climate change. For example, rainwater and greywater use systems could contribute to the efficient use of resources while sustainable drainage and the retention of surface water could facilitate groundwater recharge.
Demographic changes	Government projections indicate an increase of around 3.38 million households in England and Wales between 1996 and 2021. Most of these new households are likely to be smaller, which may increase both the overall demand for water and the amount of surface water runoff.
Reducing surface runoff and diffuse pollution	Rainwater use systems and sustainable drainage systems could facilitate the attenuation and storage of surface water runoff and potentially reduce the flood risk within a development area. Rainwater reuse systems and sustainable drainage are regarded as preventative systems, controlling both water quantity and water quality at, or close to, the source.
Potential to save costs	Sustainable water management systems designed to reduce and control surface runoff means there may be a reduced need to supplement and increase existing infrastructure to cope with increased flows. They could also lessen the need to upgrade sewage treatment works to treat increased flows as a result of surface water runoff. Rainwater and greywater use systems could assist in lowering the demand for potable water; if their implementation is widespread they could reduce the need for additional water resources and supply infrastructure. Users of rainwater/greywater use systems may also benefit from reduced water supply and wastewater bills.
Planning requirements	With the introduction of PPG25 *Development and flood risk* (DTLR, 2001a) and the amendments to Part H of the Building Regulations (DTLR, 2001b) many development plans and planning applications are encouraging or requesting the wider use of SUDS to provide the associated environmental benefits.

3   Background to rainwater and greywater use

> **Definitions**
>
> **Rainwater use system**
> A system that collects rainwater from where it falls, treats and stores it, then distributes it for use rather than allowing it to drain away. This includes water collected within the boundaries of a property, including from roofs and surrounding surfaces such as areas of hardstanding and pervious paving.
>
> **Rainwater harvesting**
> An above- or below-ground storage system that collects, treats, stores and distributes runoff of rain or snow from roofs.
>
> **Greywater use system**
> An above- or below-ground system that collects, stores, treats and disinfects greywater for use as reclaimed water in and around properties.

Superficially, the UK would appear to have plenty of water. However, the growing population, thirst for water-using appliances and climate change may be putting water resources under increasing pressure. Housing projections suggest that competition for water between the environment and people is growing, particularly in the south-east of England, which already has low water availability. Cutting demand for mains water – by using reclaimed water from rainwater and greywater use systems, for example – can help reduce our impact on the environment.

Concern is also growing about rainwater runoff from urban areas. Planning Policy Guidance Note (PPG) 25 *Development and flood risk* (DTLR, 2001a) requires local authorities to consider SUDS as means to reduce the potential for flooding and diffuse pollution affecting downstream developments. The use of rainwater close to its source within the built environment can contribute to the SUDS philosophy by helping manage surface water runoff.

DEVELOPMENT OF THE TECHNOLOGY

Rainwater use (or rainwater harvesting) systems vary in design and scale from a water butt connected to guttering and a rainwater downpipe, to larger multi-user systems with centralised collection and treatment systems. The harvesting of rainwater is a simple and relatively safe practice, which has been used for hundreds of years. Some SUDS components, such as permeable pavements, may also act as a catchment to supply water to a building.

Greywater use has developed from simply watering gardens with washing-up water or siphoned bath water during water shortages to purpose-designed packaged greywater systems that collect, treat, disinfect and distribute reclaimed water for use in buildings. Such systems are available in the UK, but have yet to be widely used.

Potential barriers to the widespread uptake of rainwater and greywater use systems within the UK include:

- the capital cost of the systems compared with the water and/or cost savings
- the cost of retrofitting rainwater and greywater use systems into existing buildings
- concern about the quality of reclaimed water
- the need for maintenance, as greywater and rainwater use systems generally are not based on fit-and-forget technology.

THE TWIN-TRACK APPROACH TO SUSTAINABLE WATER RESOURCE MANAGEMENT

Sustainable water management demands an approach that is economically, socially and environmentally acceptable and that avoids negative impacts on future generations.

The twin-track approach to addressing both water demand and supply is a holistic way of sustainably managing water resources (Defra, 2002). Many options are available.

Water demand

Future water demand is uncertain, but most forecasts predict an increase. To help manage this increasing demand a range of measures may be applied, for example:

- consumer and developer education about water efficiency
- water metering to incentivise water conservation
- the use of water-efficient appliances
- mandating water sustainability requirements in planning approval processes.

These measures can help reduce demand for water but need to be proven to be cost-effective. Most of the approaches require an initial capital investment (eg in meters or water-efficient appliances), followed by continuing costs for maintenance. In implementing these measures, it is important to raise awareness of the issues and benefits as well as to encourage positive consumer behaviour.

Water supply

The other thread of the twin-track approach is to provide sustainable water supplies. As with water demand, the future is uncertain. Exactly how will the climate change? How will technology (such as membrane technology) advance to treat previously uneconomic water resources? How will rainwater and greywater systems technology evolve?

Despite the uncertainty, the following approaches may be adopted:

- reduction in volume of water lost through leakage
- treatment and use of blackwater
- promoting infiltration to ground through SUDS
- develop new surface water resources
- abstraction from marine sources (using desalination)
- use of groundwater sources (that require higher levels of treatment) where appropriate
- recharging of aquifers at times of water surplus.

Most of the approaches to exploiting alternative water supplies are likely to require technological developments to make utilisation of these sources more cost-effective. As water demand increases and supply decreases, the business case for increased spending becomes more attractive.

The changing habits of water users that may be anticipated within the twin-track approach should help manage demand and increase awareness of measures such as using rainwater and greywater. The twin-track approach should stimulate greater uptake of these systems. Furthermore, water companies have a legal duty to promote water efficiency to their customers (a commitment overseen by Ofwat). This extends to rainwater and greywater systems as part of an overall strategy for water conservation and sustainable water use and supply.

RAINWATER AND GREYWATER SUSTAINABILITY

Rainwater and greywater use can contribute to sustainable water resource management by providing an alternative supply (rainwater) and reducing demand locally because water is used more than once (greywater). At present, rainwater and greywater systems are having only a limited impact on UK water demand and supply, because many of their benefits can only be realised when uptake is more widespread.

Rainwater and greywater can contribute to sustainability on a local scale where:

- water resources are particularly scarce
- the cost of providing a (new) mains water supply is prohibitively high
- the user wishes to reduce mains consumption because of concern over the environmental impact of their supply (abstraction from rivers or creation of reservoirs)
- the user wishes to be independent of mains water as a lifestyle choice
- the user wishes to demonstrate the feasibility of using rainwater or greywater systems.

Other sustainability issues for consideration include the use of materials and energy in the production of components and the consumption of resources during operation (for example, electricity and disinfecting chemicals).

WATER CONSERVATION IN BUILDINGS

There are five main approaches to reducing mains water use in buildings:

- water saving by good housekeeping, eg fixing dripping taps
- water-saving behaviour
- use of efficient appliances (dual- or ultra-low-flush toilets, waterless urinals)
- exploitation of alternative water supplies (use of rainwater etc)
- recycling and reuse of water (for example, greywater use).

Figure 3.1 gives a decision tree approach to assessing options available for reducing mains water use.

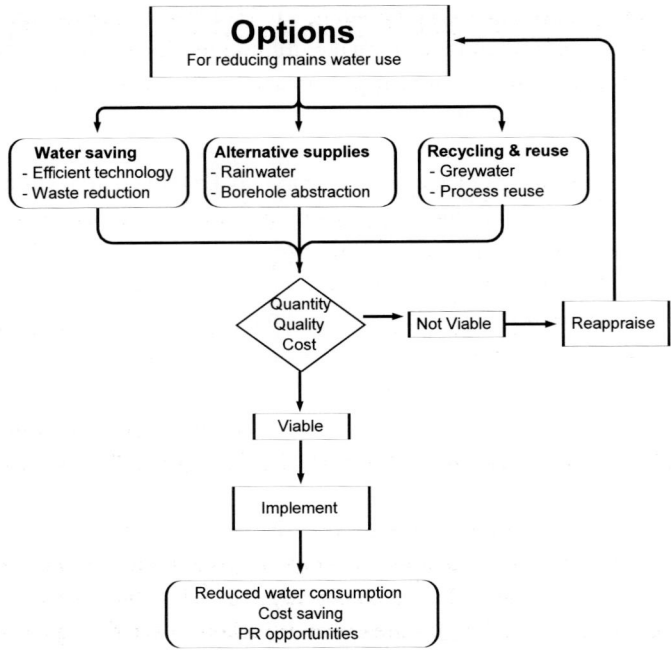

Figure 3.1 *Decision tree for reducing mains water use*

The use of water-efficient appliances, such as showers and taps, may reduce greywater arisings and also the reclaimed water yield. However, if these appliances are used to provide greywater to feed a low-volume-flush toilet, for example, then water demand should also be reduced. The ability of supply to meet demand should always be evaluated case by case.

4   Rainwater and greywater use systems

Although rainwater and greywater use systems come in different arrangements, certain components and issues are common to both. No specific legislative barriers exist to prevent the use of rainwater or greywater systems in the UK and there are currently no formal or legislative standards on water quality for rainwater or greywater systems (unless for drinking, when the Private Water Supply Regulations 1991 apply). However, the design and installation of a rainwater or greywater system in a building already being supplied with mains water will be subject to regulation (Chapter 5). It is wise to contact the local water company and or sewerage undertaker, the local building control department and also the local environmental agencies for guidance and advice.

The end use of the water determines the level of treatment or disinfection required. Using rainwater and greywater without any treatment minimises costs and reduces the consumption of materials and chemicals. This is feasible for some applications, such as garden watering, so long as measures are taken to avoid stagnation of the reclaimed water (particularly greywater) or build-up of contamination in the soil. Because of its simplicity, the use of rainwater for irrigation is commonplace and its use for toilet flushing is growing within the UK. Campaigns and subsidies by water companies have contributed significantly to the simplest form of rainwater use (water butts) and also larger-capacity systems, for example in nurseries and garden centres.

The guidance within this book focuses on the use of reclaimed water for non-drinking purposes, primarily toilet-flushing, which accounts for around one-quarter to one-third of household water use.

For detailed guidance on the design and installation of rainwater and greywater use systems readers are recommended to consult CIRIA C539 *Rainwater and greywater use in buildings – best practice guidance* (Leggett *et al*, 2001) and BSRIA TN 6/2002 *Water reclamation guidance – design and construction of systems using grey water* (Brown and Palmer, 2002a).

RAINWATER USE SYSTEMS

Rainwater can be collected or harvested from roofs and other hard surfaces around buildings. The water quality of collected rainwater depends upon the contaminants picked up from the air and the catchment area. Rainwater is generally low in contaminants so long as catchment surfaces are kept clean and systems to remove the first flush work effectively.

Rainwater use systems generally consist of one or more storage tanks, a pump, filtration units (a wide variety of specialist filters is available) and connecting pipework; some systems incorporate disinfection apparatus. There is also likely to be some form of electronic control system. In most cases there will be a connection to the mains water supply so that the system can be supplemented automatically when there is insufficient rainwater or when demand is relatively high.

Figure 4.1 relates to a typical rainwater system. The exact arrangement of these components will vary between proprietary systems.

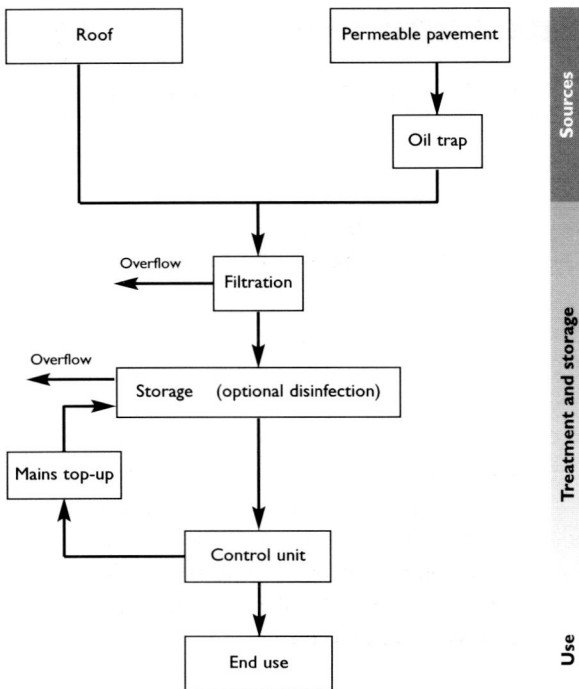

Figure 4.1 *Flow diagram of a standard rainwater system*

The system illustrated is a directly pumped arrangement, but, again, configurations vary. Some variants pump rainwater from an underground storage tank to a higher-level header tank. If insufficient rainwater is available, the tank is topped up with mains water via an AA air gap, which prevents contamination of the mains water. Some systems use a pump from the mains water tank to top up the rainwater. More sophisticated systems incorporate dual feed for rainwater and mains, which are fed directly into toilet cisterns. An air gap in the WC feed pipe prevents mains water being contaminated by backflow.

Treatment of rainwater can include filtration, biological treatment and disinfection. Filtration aims to remove solid material from the water before or after storage. An inline downpipe filter that removes debris and sediment is commonly used. If a pervious pavement forms part of rainwater use system's catchment, a suitable oil trap (oil separator) should be installed to remove oil and fuel residuals before the water is filtered.

After the water has been filtered it may need to be disinfected to kill off microbiological and bacterial contamination. Systems that use rainwater solely for toilet flushing rarely employ disinfection. This practice is well established as being safe so long as good incoming water quality is maintained and the risks of contamination from modifications to catchments and system are prevented.

Rainwater is inherently soft and, provided it is clean, may be used in washing machines (although this will require some modification to the machines), car washing and even for bathing once it has undergone appropriate treatment. Where the rainwater is to be used for drinking, washing and cooking, or used in a business that produces food or drink, the Private Water Supply Regulations 1991 apply.

The UK's weather and rainfall are becoming increasingly variable both seasonally and geographically. The demand for water tends to be relatively constant, however. Although it is possible to design and construct a rainwater system to meet 100 per cent of water requirements, it is rarely economic to do so where mains water is already provided. The cost is inflated by the need for a large collection tank (or small reservoir) and the space for a large system. It is standard practice in Germany to design the collection tank so that it overflows about twice a year to allow surface debris in the tank to be cleared. This is achieved by careful modelling of the tank size on an individual system basis.

A rainwater system is more likely to be optimised to provide useful savings of mains water at a reasonable cost. This assessment will take into account factors such as ready access to the available catchment surfaces, tank size and location, water quality requirements and potential usage.

GREYWATER USE SYSTEMS

Sources of greywater include sinks, baths, showers or clothes-washing machines. The water quality will depend on the contaminants picked up during the use of the water. Greywater is generally warm, nutrient-rich and high in contaminants, making it an ideal medium for microbiological activity and bacteriological growth. The majority of packaged greywater systems available for the domestic market are not designed to use kitchen sink or dishwasher wastewater.

Greywater derived from baths, showers, hand basins and washing machines is normally less contaminated than greywater from kitchen sinks and dishwashers. The latter two sources may have a high fat, oil and foodstuff content, but the former are likely to contain pathogens. As a result, most packaged greywater systems available for the domestic market are designed to use only greywater derived from personal cleaning.

Greywater systems generally consist of one or more storage tanks, a pump, filtration units, chemical dosing (for disinfection) and connecting pipework. There is also likely to be some form of electronic control. In most cases there will be a connection to the mains water supply so that the system can be automatically supplemented when greywater arisings are low or demand is relatively high. Figure 4.2 relates to a typical greywater system. Proprietary systems differ in how they arrange these components and the technology employed for filtration and disinfection.

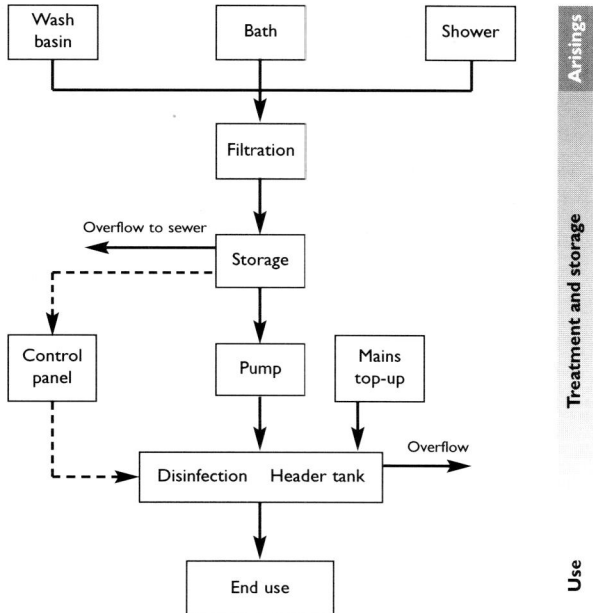

Figure 4.2 *Flow diagram of a typical greywater system*

Figure 4.2 is based on a greywater system that uses a header tank in the roof space; as with the rainwater system, there are several variants. For example, some direct-fed systems have the storage treatment and control panel in a modular unit able to provide reclaimed water for WC flushing.

The most common use of greywater is for toilet flushing. In part this reflects the water quality and the potential risks of using untreated greywater, and also the close volumetric match between the demand for toilet-flushing water and greywater arising in domestic situations.

It is important to coarse-filter greywater before storing it. This is to prevent the build-up of debris in the storage tank, which would encourage bacterial growth. Self-cleaning filters are preferred, because they reduce the need for user intervention or maintenance. Some communal systems have incorporated sand and carbon filters and others have used membrane filters. Although they are relatively more expensive to apply, these filters provide a consistent end result.

Some systems incorporate biological treatment that encourage the development of a biofilm or bacterial colonies that breakdown nutrients and contaminants. Normally chemical disinfection is used, although ultraviolet disinfection can also be used if the water quality permits (this technique requires low levels of turbidity).

COMBINING RAINWATER SYSTEMS AND SUDS

Rainwater harvesting systems may be used to reduce the volume of runoff from a building or development, enabling surface water to be controlled at source. This can provide a source control function and is often considered as a prevention or management technique. However, it will only provide a temporary source control function until the storage tank is full, as the tank will overflow to a drainage system.

Stored water is held in storage tanks and the permanent storage volume required for reuse is provided in addition to the volume required to attenuate stormwater flows, unless a continuous rate of use can be guaranteed.

5   Policy, regulation and guidance

There is no formal UK Government policy on the operation of rainwater and greywater use systems either in the home or within an industrial development. However, through the promotion of sustainable development and other initiatives, the government encourages a reduction in the consumption of resources. Uptake of sustainable water management systems would go a long way towards achieving this goal.

REGULATIONS

Anyone considering installing a rainwater or greywater system should consult their local water company to ensure it will comply with the current regulations. Although there are no legislative barriers to prevent the use of rainwater or greywater systems in the UK, numerous regulations (listed below) affect their installation and use. Some, like the Health and Safety at Work Act, are relevant to systems even when installed in a private house.

Health and Safety at Work, etc Act 1974

This applies to rainwater/greywater use systems where electrical or potentially hazardous chemicals are used. Where chemicals are used for disinfection in non-domestic installations they would be subject to COSHH as well. The Health and Safety at Work Act contains a requirement to minimise the risk of exposure to biological hazards, which could include bacteria and coliforms.

Water Industry Act 1999

The Act permits sewerage undertakers to levy additional charges to customers who discharge from a source other than a potable network into a foul sewer following use in the home. In practice, water companies often ignore the issue of discharge of rainwater from roofs into foul sewers because the volumes are relatively small. In some instances, however, this may cause problems with the drainage infrastructure.

Water Supply (Water Fittings) Regulations 1999

These Regulations are concerned chiefly with preventing contamination of the mains water supply and reducing wastage. With regard to the installation and operation of rainwater/greywater use systems the main issue is prevention of cross-contamination and backflow, primarily through the specification of an air gap.

The Regulations stipulate that all pipelines be identified, including those carrying greywater and those providing reclaimed water for end use (guided by WRAS). They also provide a specifications designed to ensure efficient use for WC flush volumes and other water-using appliances.

Building Regulations 2000 (as amended)

These relate to greywater systems and the connection of sinks, showers and baths (and appliances) providing greywater to a vented stack. This may prevent loss of the water seal in the traps associated with individual fittings or appliances.

Private Water Supplies Regulations 1991

Contain information on the collection of water for human consumption.

Legionnaires' disease. Approved code of practice and guidance

The ACOP for the control of the *Legionella pneumophila* bacterium in water systems (HSC, 2000).

Construction (Design and Management) Regulations 1994, amended 2000

The CDM Regulations set down standards for health and safety within project planning and management of construction and maintenance. Systems should be designed and installed to facilitate safe operation and management.

Control of Substances Hazardous to Health Regulations 1999

The COSHH Regulations cover the use of chemical disinfection and microbiological hazards that may arise during maintenance.

The Electricity at Work Regulations 1989

These are relevant to the installation of rainwater and greywater systems in non-domestic applications, primarily for the use of a control system and pumps.

Personal Protective Equipment at Work Regulations 1992

The Regulations concern the provision of PPE, such as eye and hearing protection, masks, respirators and gloves.

Provision and Use of Work Equipment Regulations 1998

PUWER introduces a requirement to report injuries and dangerous occurrences.

Confined Spaces Regulations 1997

These require a risk assessment and safe method of work for unavoidable entry to confined spaces, such as the cleaning of large tanks. On commercial or industrial sites a permit to work will normally be required.

Manual Handling Operations Regulations 1992

The aim of these Regulations is to reduce the incidence of injury and ill-health arising from the manual handling of loads at work.

Workplace (Health, Safety and Welfare) Regulations 1992

For rainwater and greywater use systems, the Regulations are of importance in setting down employers' responsibilities to ensure that the systems can be operated and maintained safely. They require a risk assessment to be undertaken to identify potential impacts on persons coming into contact with reclaimed water.

GUIDANCE

WRAS guidance

The Water Regulations Advisory Scheme (WRAS) gives advice on the Water Supply (Water Fittings) Regulations 1999. This is in the form of information and guidance notes covering the specification and design of reclaimed water systems. Guidance notes can be downloaded from WRAS at <www.wras.co.uk>.

WRAS guidance on reclaimed water systems

Guidance note 9-02-04 *Reclaimed water systems* (WRAS, 1999a) aims to support water conservation and prevent cross-contamination of mains water from reclaimed water reclamation systems. It includes information about the causes of contamination, classes of water quality and useful information on the installation, operation and management of reclaimed water systems (including rainwater and greywater use systems). The note also provides a hazard assessment methodology based on the type of systems and potential exposure.

WRAS guidance on marking and identification of pipework

Guidance note 9-02-05 *Marking and identification of pipework for reclaimed (greywater) systems* (WRAS, 1999b) stipulates that pipes carrying reclaimed water should be clearly distinguishable from those carrying mains water, to avoid accidental cross-connection.

Pipework identification guidance

The Water Fittings Regulations 1999 require pipework used for rainwater and greywater use to be identified. All pipework should be labelled in accordance with BS 1710:1984 and the guidance in WRAS 9-02-05 (WRAS, 1999b). Care must be taken not to cross-connect reclaimed water and mains water pipework during installation or subsequent work on a property. Pipe marking is essential to help prevent accidental cross-connections, which could contaminate mains water supply.

WRAS guidance on pipe marking is:

Greywater	green/black/green bands with the word GREYWATER in black lettering on a grey background
Reclaimed water	green/black/green bands with an additional white band in the centre and the words RECLAIMED WATER in black lettering on a green background (as below).

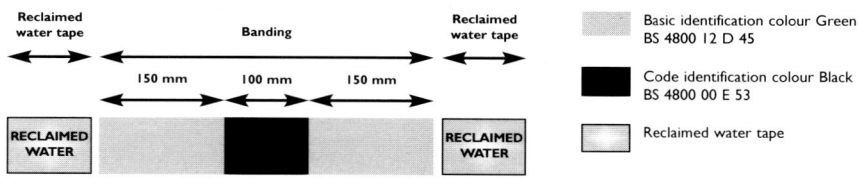

Figure 5.1 *WRAS colour coding of pipework*

Outlets to supply reclaimed water must also be clearly labelled according to quality and end use (WRAS, 1999b).

Dealing with risk

The level of risk that exists when using reclaimed water varies with the quality of the rainwater or greywater that is collected and the end use. During the design of a system the quality of the raw water, the end use and the characteristics of the end-users should be fed in to a risk assessment to determine the most appropriate treatment methods, the ongoing monitoring schedule and the frequency of maintenance.

The scale of the system should also be considered. Multi-user systems can represent a higher risk, because the number of users exposed to the reclaimed water is greater. Users may include the very young or very old, who are more susceptible to illnesses. The ability to control the quality of the inflowing rainwater and/or greywater may also be diminished in larger-scale systems, such as educational establishments or halls of residence. Definitions of single and multi-user systems are given in Table 5.1.

Table 5.1. *Definitions of single and multi-user systems*

Single-user system	Rainwater and greywater use system designed for a single user, normally a household
Multi-user system	Residential and commercial rainwater/greywater use systems designed to serve multiple properties and users, such as flats, residential and/or housing associations, hotels, offices and other commercial applications

Conversely, larger-scale systems and the associated economies of scale may facilitate improved maintenance procedures, which will reduce the potential risk of failure and also limit users' exposure to the mechanics of a system for an individual household. Risk assessments should be undertaken on a case-by-case basis. The box below provides more information about risk assessments.

Risk assessment

A risk assessment should consider factors such as:

- the source of the raw water, ie
  - condition or age of the roof (rainwater use systems)
  - potential sources of contamination
  - source of the greywater
- number and type of users (ie school, hospital, commercial buildings)
- exposure to risks
- users' awareness of the reclaimed water
- potential end use.

General information about risk assessments can be found on the HSE website <www.hse.gov.uk>.

Levels of risk are likely to be low for rainwater use systems. Rainwater is invariably less polluted than greywater, although it may be contaminated by bird droppings, leaves or other organic matter from the catchment area. In the single-user environment, the rainwater system user is likely to be aware of the risks and can take appropriate steps. In this case, treatment, beyond settlement and coarse filtration, is unlikely to be required. Users in a multi-user environment may not be so aware of the nature of the reclaimed water. Because the risks from cross-contamination or incorrect use of the reclaimed water are greater, some treatment may be justified.

Levels of risk to the end-user of a greywater use system are higher than for a rainwater system. Greywater contaminants include fats, grease, cleaning products and in some cases human faecal matter. Treatment (including disinfection) of the greywater is needed to ensure that the reclaimed water is safe for its intended application.

Outputs from the Buildings That Save Water project (Leggett *et al*, 2001a and b; Brewer *et al*, 2001) give more information about risks and hazards and operability (HAZOP) studies on greywater systems. So far, the emphasis of research has been on greywater systems, the safety of which is chiefly determined by the proper functioning of the disinfection systems and control features.

Water quality guidance

Some technical guidance documents relate to the quality of reclaimed water, but there is no legally binding standard within the UK.

Guidelines and recommendations for the use of reclaimed water in the UK have been produced by:

- CIRIA: Buildings That Save Water (Leggett *et al*, 2001a)

- BSRIA: *Water reclamation guidance* (Brown and Palmer, 2002a)

- WRAS: *Reclaimed water systems* (WRAS, 1999a).

The American USEPA guidelines for water reuse (USEPA, 1992) are also used sometimes and are summarised in Table 5.2. Following these guidelines should help reduce the risk of disease transfer from the use of rainwater or greywater. Different countries have different requirements and there is no universal agreement on the parameters to be measured, although concerns commonly relate to microbial activity and nutrient levels within reclaimed water.

The risks involved with these systems are described below, followed by some guidance on the appropriate treatment methods for different systems.

Table 5.2   *Summary of water quality technical guidance for non-potable water*

Parameter	BSRIA*	USEPA¤	WRAS⊘	Bathing Water Directive
BOD_5 *	Dissolved oxygen > 1 mg/l	< 10 mg/l BOD_5	Not provided	Not provided
Turbidity ◊	Opacity < 60 per cent At 254 nm	< 2 NTU	Not provided	Not provided
Microbiological quality •	< 1000 cfu/100 ml total coliforms	(*E.coli*) < 1 cfu/ 100ml	< 10 000 cfu/100 ml faecal coliforms < 100 cfu/100 ml faecal enterococci	< 10 000 cfu/100 ml total coliforms < 2000 cfu/100 ml faecal coliforms
pH (measure of acidity or alkalinity)	Not provided	6–9	Not provided	Not provided
Chlorine (Cl_2) residual disinfectant	0.5–2 mg/l	< 1 mg/l	Not provided	Not provided

Notes

* BSRIA guidance (Brown and Palmer, 2002a) is for toilet flushing with greywater only (details in Table 5.5).

¤ USEPA, 1992.

⊘ WRAS, 1999a.

◊ The BSRIA test for opacity is simpler than USEPA's turbidity test.

• Various standards are recommended. The test for total coliforms is simplest, but still would have to be undertaken by a laboratory.

TREATMENT RECOMMENDATIONS FOR RAINWATER USE SYSTEMS

Treatment recommendations for rainwater use systems are set out in Table 5.3. The chosen treatment should be subjected to a risk assessment as described above to determine whether disinfection of the rainwater is required.

Table 5.3 *Design recommendations for rainwater use systems*

Application	End use	Treatment requirements
Single-user	WC flushing, irrigation and other non-potable uses	Settlement and coarse filtration
Multi-user	WC flushing, irrigation and other non-potable uses	Settlement and coarse filtration (plus disinfection to achieve total coliforms < 1000 cfu/100 ml if thought necessary)

Testing regime for rainwater use systems

If disinfection is needed for a rainwater use system, testing should be carried out as part of the maintenance programme to ensure that the reclaimed water meets the required standard. These requirements are described in Table 5.4 below.

Table 5.4 *Testing recommendations for rainwater use systems that use disinfection*

Test	Value	Frequency	Location
Residual disinfectant	> 0.5 mg/l (chlorine)	Annual	Point of use (eg WC cistern)
Or *			
Total coliforms	< 1000 cfu/100 ml	Annual	Point of use (eg WC cistern)

Notes

* A test for either residual disinfectant or total coliforms is required, because if residual disinfectant is present there should be no coliforms risk

Note that only one of the above tests is required. Usually this would be the test for residual disinfectant. This can be carried out using *in-situ* hand-held monitoring equipment. Tests for total coliforms need to be carried out under laboratory conditions. Where another method of disinfection is used (such as bromine) an appropriate residual level of disinfectant should be applied in accordance with the manufacturer's instructions.

It is unlikely to be beneficial to test rainwater systems more than once a year. Testing frequency for large multi-user applications may be increased to six-monthly intervals. Tests should be carried out using water at the point of use, such as the WC cistern.

Any visual deterioration indicates a reduction in the quality of the reclaimed water, which may necessitate an inspection and remedial maintenance work.

TREATMENT RECOMMENDATIONS FOR GREYWATER USE SYSTEMS

This section sets out the treatment requirements for greywater use systems. More tests are required than for rainwater use systems to ensure that the water is of the appropriate quality for the intended use. There are no legally binding water quality standards for reclaimed water. In their absence, the most appropriate form of guidance is the BSRIA standard set out in Table 5.5.

Type approval

BSRIA's water reclamation guidance sets out a standard for the benefit of manufacturers, specifiers, installers and end-users of water reclamation systems. The standard provides a methodology for establishing the safety and performance of packaged water reclamation systems.

Systems that comply with the standard are expected to function safely and achieve the required reclaimed water quality in circumstances specified by the system manufacturer. The performance should be verified during the commissioning phase before reclaimed water is used and thereafter through periodic testing. Type approval of greywater use systems should remove the need for complex testing against each of the individual parameters. Instead, a simple test of the level of residual disinfectant during each maintenance visit should be carried out. Additional tests may still be required for larger multi-user greywater systems.

Treatment design recommendation

The BSRIA standard applies only to reclaimed water for three types of end use, namely:

- toilet flushing
- gravity-fed irrigation systems
- pressure applications (eg vehicle washing).

Pressure applications require higher-quality water than is appropriate for WC flushing. This is mainly related to the greater risk of exposure and ingestion through the formation of spray and aerosol. A summary of these standards are set out in Table 5.5.

Table 5.5 *Design guidelines for greywater use systems (Brown and Palmer, 2002)*

Use	Test	Value
WC flushing	Total coliforms	< 1000 cfu/100 ml
	Max residual disinfectant *	2 mg/l (Cl_2)
	Min residual disinfectant	0.5 mg/l (Cl_2)
Gravity-fed irrigation systems	Total coliforms	< 1000 cfu/100 ml
	Max residual disinfectant *	0.5 mg/l (Cl_2)
	Min residual disinfectant	N/A
Pressure applications	Total coliforms	10 cfu/100 ml
	Max residual disinfectant *	0.5 mg/l (Cl_2)
	Min residual disinfectant	N/A
All applications	Dissolved oxygen	Dissolved oxygen > 1 mg/l
	Turbidity ◊	Opacity < 60 per cent at 254 nm

Notes

* For disinfectants other than chlorine appropriate residual concentrations and appropriate analytical methods should be in accordance with the manufacturer's recommendations.

◊ Normally measured with a 1 cm cube

Testing regime for greywater use systems

The testing regime required depends on whether the greywater use system has type approval and on the number of users. The risk associated with greywater systems increases with the number of users because of factors such as cross-contamination and the discharge of unsuitable substances to the greywater equipment.

As with rainwater use systems, the economies of scale associated with larger greywater systems may facilitate improved maintenance procedures, thereby reducing potential risk of failure. Users' exposure to the mechanics of a system may also be reduced relative to that for an individual household. Where greywater systems do not have type approval, risk assessments should be undertaken on case by case.

Tables 5.6–5.9 set out the proposed testing requirements for greywater for both single-user and multi-user environments for equipment with and without type approval. The required treatment values should be in accordance with the guidelines in Table 5.5.

There is an option for measuring residual disinfectant or total coliforms. Usually this would be the test for residual disinfectant, as *in-situ* hand-held monitoring equipment can be used. Tests for total coliforms need to be carried out under laboratory conditions. Where another form of disinfection is used (such as bromine) an appropriate residual level of disinfectant should be applied in line with the manufacturer's instructions. All samples should be taken from the point of use, for example the WC cistern.

Any aesthetic deterioration in the water (bad odour or poor visual appearance) indicates a reduction in the quality of the reclaimed water. An inspection and remedial maintenance work may be required.

Table 5.6   *Testing recommendations for greywater use systems: single-user with type approval*

Test	Frequency	Value
Residual disinfectant	Annual	See Table 5.5
Or *		
Total coliforms	Annual	See Table 5.5

Notes

* A test for either residual disinfectant or total coliforms is required, because if residual disinfectant is present there should be no coliforms risk.

Additional annual tests may be required for greywater equipment that does not have type approval to ensure that it conforms to the BSRIA guidelines (set out in Table 5.7).

Table 5.7   *Testing recommendations for greywater use systems: single-user without type approval*

Test	Frequency	Value
Dissolved oxygen	Annual	Dissolved oxygen > 1 mg/l
Turbidity *	Annual	Opacity < 60% at 254 nm
And		
Residual disinfectant	Annual	See Table 5.5
Or ◊		
Total coliforms	Annual	See Table 5.5

Notes

* Normally measured with a 1 cm cube

◊ A test for either residual disinfectant or total coliforms is required, because if residual disinfectant is present there should be no coliforms risk

Testing of multi-user systems is likely to be more frequent than for single-user systems, and this should be considered in the risk assessment. Tables 5.8 and 5.9 set out the recommended frequency of testing.

Table 5.8 *Testing recommendations for greywater use systems: multi-user with type approval*

Test	Frequency	Value
Dissolved oxygen	Annual	Dissolved oxygen > 1 mg/l
Turbidity *	Annual	Opacity < 60% at 254 nm
And		
Residual disinfectant	Every 3–6 months	See Table 5.5
Or ◊		
Total coliforms	Every 3–6 months	See Table 5.5

Notes

* Normally measured with a 1 cm cube

◊ A test for either residual disinfectant or total coliforms is required, because if residual disinfectant is present there should be no coliforms risk

Table 5.9 *Testing recommendations for greywater use systems: multi-user without type approval*

Test	Frequency	Value
Dissolved oxygen	Six-monthly	Dissolved oxygen > 1 mg/l
Turbidity *	Six-monthly	Opacity < 60% at 254 nm
And		
Residual disinfectant	Monthly	See Table 5.5
Or ◊		
Total coliforms	Monthly	See Table 5.5

Notes

* Normally measured with a 1 cm cube

◊ A test for either residual disinfectant or total coliforms is required, because if residual disinfectant is present there should be no coliforms risk

6   Maintenance

The impact and sustained operation of rainwater and greywater use systems depend on the enthusiasm and commitment of either the users or those responsible for maintenance. Because the technology requires some user intervention, a lack of commitment in undertaking maintenance may cause problems for system users and lead to benefits not being realised.

Before the maintenance regime for a rainwater or greywater use system is implemented the system should be inspected at handover to ensure that the client has a robust and operational system that is unlikely to fail as a result of errors in design and/or installation. The handover process should include a manual that incorporates details on servicing and maintenance requirements for the specific system.

The enclosed model agreement is a legal document that can be completed to form the basis of an agreement between two parties regarding the maintenance and operation of sustainable water management systems. This should help ensure that systems operate as designed and that routine maintenance requirements are undertaken.

If the reader is considering purchasing a greywater use system it may be helpful to check that it complies with BSRIA's standard for water reclamation. In other situations, the manufacturer will need to provide the information detailed in Table 6.1.

Table 6.1   *Information to be supplied by systems manufacturer*

Contact details	Name of manufacturer, address, telephone and fax number, email and web address
System details	Model number/serial number, date of manufacture and place of purchase
	A list of system parts and specifications, and a separate list of consumables such as disinfectant and rubber seals. Possibly in graphical format with each part given a number and listed
Installation instructions	The operation of the unit should be explained (ie what it does and how it works)
	A list of additional items required for installation (including specialised tools) that are not included with the system
	Installation instructions should include safety considerations. They should be clear, concise and listed in order. Diagrams should be used
Commissioning instructions	This should contain a list of checks to be made before the system is activated. A tick box checklist might be helpful here
	A fault-finding procedure, possibly in a table or flow chart format
User instructions	Instructions on general day-to-day system operation, which should include simple tasks that the user is expected to do, and a schedule setting out when trained maintenance personnel are required. A simple fault-finding guide for users
Operation and maintenance instructions	In-depth instructions for on safe system maintenance, including in-depth system fault-finding
Maintenance log book	This should be used to record details of maintenance carried out
Other information	A list of product approvals
	Certification of warranty and conditions if given
	A list of recognised service agents or retailers of the system

CONSIDERATIONS DURING MAINTENANCE

Well-designed rainwater and greywater use systems should be safe to operate. However, installation and maintenance activities may create risks. The risks are mainly associated with greywater use systems. Some of these risks may be exacerbated during periods of system failure. Those undertaking maintenance should carry out risk assessments and should follow procedures to minimise risks. The main areas of concern (adapted from BSRIA's guidance) are set out below.

Accident hazards

Care must be taken when inspecting and working on elements of systems. This could include inspection of storage or header tanks, working on roofs and gutters and general grounds work.

Biological hazards

It should be assumed that any part of the system may be contaminated with pathogenic bacteria (this is particularly relevant to greywater use systems). Maintenance personnel and others who come into contact with greywater or reclaimed water from rainwater and greywater use systems should wash their hands before eating and wash overalls after use. Greywater collection pipework and tanks should be treated as if contaminated with faecal material: operatives should wear gloves and overalls when cutting into systems and when cleaning tanks.

Chemical hazards

In undiluted form the chemicals used for disinfection may be hazardous through skin contact, ingestion or inhalation. Containers must be properly labelled and kept in a secure place out of reach from of children. Suppliers must give guidance on handling precautions and instructions on what to do in the case of accidental skin contact, ingestion or spillage.

INSPECTIONS

Routine inspections should be carried out in accordance with manufacturer's instructions and guidance on testing (this should be included in the model agreement). Annual testing is suggested for single domestic rainwater use systems, and six-monthly to annual testing is recommended for greywater use systems.

For multi-user installations this frequency should increase in line with the perceived risk. Rainwater use systems should be tested six-monthly and greywater use systems every month. More frequent inspections and testing may be required during commissioning and in the early phases of operation, or where the impacts of failure are high.

Regulators alone cannot enforce water quality monitoring effectively. However, the model agreement can make monitoring by the maintainer a contractual obligation. Monitoring is considered to be prudent, particularly where many people are using the water. Reducing risks needs to be addressed in system design, however. Systems should be able to meet high water quality criteria and be designed to fail-safe if any component of the treatment system fails.

System failure

Proprietary rainwater systems and rainwater harvesting systems tend to be reliable and manufacturers generally provide instructions on system use and fault-finding. Manufacturers are being encouraged to incorporate a control panel to provide users with information on water levels and the operation of individual components such as pumps. Pump failure is the most common cause of unplanned maintenance within rainwater use systems.

Greywater systems are more complex than rainwater systems, as they have a greater number of components and more sophisticated treatment processes. Users should be given instructions on operation and fault-finding. Because the failure of a rainwater or greywater treatment system may create health risks, many systems have a fail-safe mechanism to prevent untreated water being supplied for use. Failures can include:

- failed or faulty pumps
- control panel failure
- disinfectant inadequate or expired
- faulty switch valves
- filter blockage.

Ensuring correct installation and commissioning contributes to system reliability, but during the handover stage users should nevertheless be instructed on fault diagnosis. Information should cover pipework, integrity of storage and the reliability of components such as pipes and switch valves.

MAINTENANCE

System design should facilitate safe and convenient access by personnel and, where necessary, construction plant. There may need to be periodic access to collection tanks so that pumps can be maintained and filters changed or cleaned; occasionally tanks themselves may need to be cleaned.

For greywater systems, header tanks may also need to be accessed to maintain level controls, valves and control systems.

Where the equipment supplier does not undertake maintenance, it should supply full instructions to the customer, including:

- the name, address and telephone number of the system supplier and/or delegated maintenance company
- the model and serial/contract number of the system
- if appropriate, the approvals body (and certificate number) certifying compliance with standards; in the case of greywater it is BSRIA's *Water reclamation guidance* (Brown and Palmer, 2002a)
- general safety advice including the avoidance of hazards
- guidance on the effective use of the system to maximise the sustainable management and use of water
- start-up and shut-down procedures
- periodic checks to be carried out by the operator
- planned monitoring and maintenance requirements
- consumable requirements
- basic fault-finding techniques.

Routine maintenance

The majority of systems on the market are not "fit-and-forget", as they require periodic checking and maintenance to ensure trouble-free operation. Where systems are well maintained their reliability is considerably enhanced. Some activities will be suitable (although not necessarily desirable) for most users to undertake, for example replacing or cleaning filters and topping-up disinfectant reservoirs. Other tasks, such as servicing the submerged pump or control panel, may require specialised service visits.

There are wide differences in the extent of planned maintenance between systems. Single domestic and some multi-user rainwater systems are essentially maintenance-free apart from the need to clean gutters and/or catchments and filters. Some greywater system manufacturers are developing systems intended to be maintenance-free, whereas others require three-monthly inspection and disinfectant top-up.

The frequencies indicated in Tables 6.2 and 6.3 are based on findings from the BTSW project. In the first instance, users should contact the system or component manufacturer for advice on the frequency of maintenance and monitoring.

Rainwater systems

Many systems are designed to reduce the level of user intervention required. As such, systems are simple to maintain, but periodic checks may be needed. The main components that may require routine operation and maintenance checks are the pump, valves, filters and disinfection system (if installed). The frequency for maintenance activities suggested in Table 6.2 is based on a single domestic system; for multi-user applications more frequent maintenance visits may be needed.

Table 6.2 *Typical frequency of maintenance activities for rainwater systems*

Component of system	Maintenance frequency
Manually cleaned filters	Clean monthly
Self-cleaning and/or coarse filters	Check and clean every three months depending on site (eg the degree of tree cover)
Roofs and gutter	Clean once or twice a year, depending on site
Ultraviolet disinfection	Six-monthly or annual replacement, depending on system
Chemical disinfection	Some systems require monthly replacement of disinfectant
Pump	Annual check of functioning and wiring
Tank	Visual inspection of the tank is recommended at least once a year. Excessive silt should be removed.
	Cleaning and draining down the tank may not be required for some time (every 10 years)
Mains water top-up	Check every six months to annually to ensure it is working

Pumps are reliable so long as they run within the design specification. Pumps on small domestic systems are designed to be maintenance-free. Larger pumps may require periodic maintenance, greasing, seal changes etc.

Greywater systems

Designs are increasingly aimed at minimal user intervention, but most systems require disinfectant to be added and possibly filters to be checked. Monitoring is also necessary in some systems to ensure they are operational and providing savings. The frequency of disinfectant top-up can vary from three to 12 months, depending on the design. Most systems prevent untreated greywater being used when disinfectant runs out.

Experience from the BTSW projects suggested that where there was no clear indication of system status on the control panel, users were often completely unaware of a failure and the system would automatically use mains water.

Table 6.3 *Typical frequency of maintenance activities for greywater systems*

Component of system	Maintenance frequency
Manually cleaned filters	Clean monthly, but this depends on the system and the sediment loading of the water
Self-cleaning and/or coarse filters	Check and clean every three months, depending on site (eg the degree of tree cover)
Biological cultures	Periodic cleaning – frequency and replacement of sand or media are dependent on system and should be in line with manufacturer's specification
Ultraviolet disinfection	Six months or annual replacement depending on the system
Chemical disinfection	Some systems require monthly replacement of disinfectant
Membrane treatment	Periodic cleaning – frequency and chemical dosage in line with manufacturer's specification
Pump	Annual check of functioning and wiring
Tank	Visual inspection of the tank is recommended at least once a year. Excessive silt should be removed.
	Cleaning and draining down the tank may not be required for some time (every 10 years)
Mains water top-up	Should be checked annually to ensure it is working

Operation and maintenance manuals for greywater systems should be supplied with all systems. This literature should include system specification and technical information together with more detailed maintenance and servicing requirements.

Manufacturers of packaged domestic greywater systems recommend an annual service by a qualified person (although users may prefer more frequent checks). This would mainly entail the cleaning of filters and functional system checks. There may also be recommendations for other maintenance tasks on a longer time-scale, eg checking the electrical integrity of the pump.

For multi-user systems, operation and maintenance of the system would normally lie with the building owner or operator. Operational checks and maintenance of system components may have to be undertaken more regularly and by professional staff. These requirements can be accommodated under the model agreement. However, it is possible that basic monitoring and replenishment of consumables may be undertaken by appropriately trained on-site staff or contracted out (also under the model agreement).

Waste management

Filters that have removed or replaced should be handled with care and disposed of safely. Disposable filters from domestic systems may be included in general household refuse. Ultraviolet lamps use mercury vapour, so should not be disposed of in general refuse. Local authorities should be contacted for advice on the appropriate disposal of filters from multi-user systems and UV lamps.

Correct disposal of waste from the cleaning of storage tanks is essential and may include use of a waste product cleansing service, in which case a specialised contractor may be required. Underground storage tanks may also need to be cleaned by using pump systems or by entering the tank. Care should be taken when working in confined spaces.

Handover of properties

Prospective purchasers of a property must be made aware that a rainwater or greywater system is installed and the system's maintenance and operational requirements should be stated clearly to the buyer.

7 Development of rainwater and greywater use systems model agreement

The model agreement developed for this project was based on a detailed legislation review and consultation exercise. The review gave the legal framework for the agreement and the consultation provided a list of potential scenarios for which the agreement would need to be used. The model agreement is based on current legislation (mid-2004) and can be found enclosed with this guide. It is also available electronically on the enclosed CD-ROM and can be downloaded from the CIRIA website at <www.ciria.org/suds>.

THE MODEL AGREEMENT

The model agreement (RW/GW MA) is a simple contract between the property-owner or tenant ("the customer") and the maintenance provider ("the maintainer"). Its main purpose is to facilitate continuing maintenance of rainwater and greywater use systems. The owner may be either a single household or a multi-user organisation such as a residential establishment, a housing association or a commercial body. This contract sets out the responsibilities of the parties, the number of maintenance visits and the charges for the services. The main elements of the agreement are shown in Figure 7.1.

The contract has two main components:

- the model agreement
- the schedule.

The model agreement places obligations on both the maintainer and the customer. The maintainer must ensure that all maintenance duties are carried out effectively, as instructed in the schedule of work. The customer, for example, has to give the maintainer access to the system.

Details of the services to be provided are set out in the schedule. This can be amended to provide specific information about the maintenance tasks required. It should specify the activities that ought to be undertaken on each maintenance visit. Commentary and guidance on the model agreement is provided in Chapter 8.

The agreement is set out in accordance with the legislative review, one of the key areas being water quality. The BSRIA standards for reclaimed water (Brown and Palmer, 2002a) are recommended for the quality of the greywater. No treatment would be expected for rainwater systems, except when deemed necessary by the risk assessment.

Scenarios for the model agreement

The model agreement for rainwater and greywater use is suitable for scenarios that range from a single domestic user to multi-user situations. In the case of a single user the agreement will often be between the householder and the maintenance organisation. For some larger single systems the agreement may be between a building operator or facilities management company and the maintainer. An additional agreement may be needed between the tenant and/or property-owner and the building operator or facilities management company. This is represented in Figure 7.1.

The multi-user agreement will often be for a communal system serving a housing association, industrial estate or commercial business park, for example. The parties to the agreement may include a facilities manager and/or building operator and there will often need to be an additional agreement between those that benefit from the system and the organisation that manages the building or site.

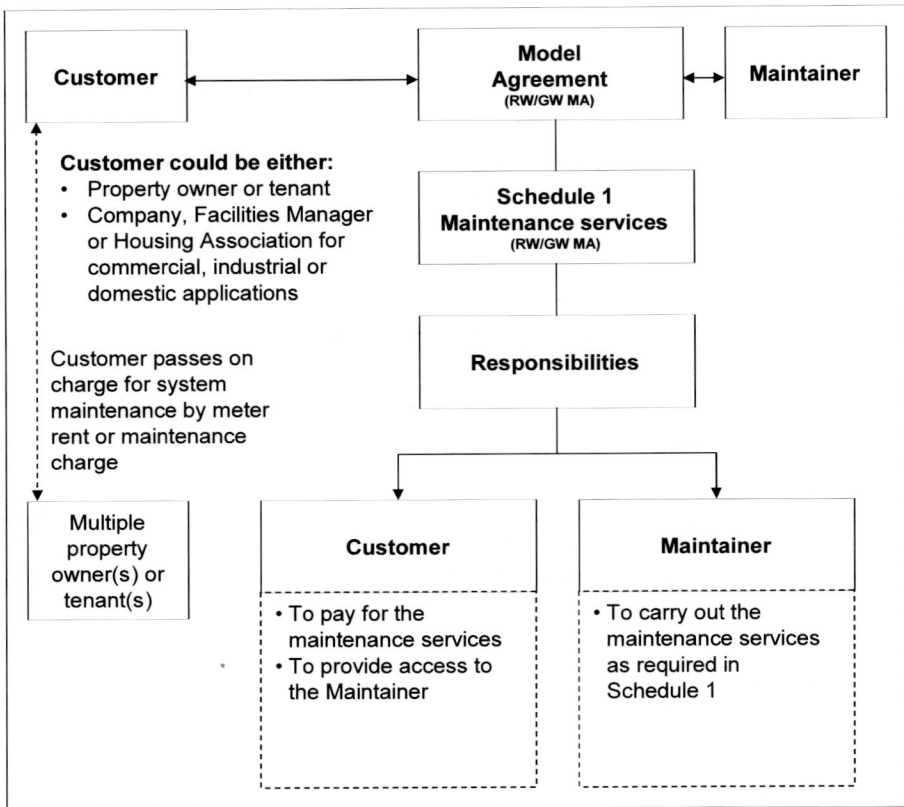

Figure 7.1   *Mechanism for the rainwater and greywater use model agreement*

Model agreements and the planning framework

Where water resources are scarce, the local authority may insist that rainwater or greywater use systems are incorporated into developments. It can do this by developing an agreement under the Town and Country Planning Act 1990, which sets out the framework for planning and development control in England and Wales.

The 1990 Act allows local planning authorities to attach conditions when granting planning permission for developments and these can be used to encourage sustainable water management.

Section 106 of the Act allows local planning authority to enter into a legally binding agreement with a third party or parties – commonly referred to as a "planning obligation". The Section 106 planning obligation can:

- restrict the level and type of development

- require specific activities to be carried out in, on, under or over the land

- require money for specified purposes to be paid to the local planning authority (on a specified date or dates or periodically).

These criteria allow the local planning authority to specify, within the obligation, the work to be carried out in connection with the sustainable water management. This is underpinned by the Local Government Act 2000, which charges local authorities with duties to promote the improvement of the environment and to contribute to sustainable development of their areas.

Section 106 agreements can be used to specify the infrastructure to be used within developments, particularly water supply and wastewater infrastructure such as rainwater and greywater use systems. The model agreement provided with this book can be used within the S106 framework to ensure that the systems are maintained, helping to sustain potential water savings and to minimise health risks.

8   Commentary on the model agreement

COMMENTARY ON THE MODEL AGREEMENT

The following section sets out guidance to the model agreement for rainwater and greywater use systems (RW/GW MA).

Section 1 – Details of parties	
	The details of both parties (maintainer and customer) should be entered into the tables.
	This defines the two parties to the agreement, who are from here on referred to as the "maintainer" and the "customer".

Section 2 – The agreement	
	This section sets out the charges agreed for the scheduled maintenance visits and additional services. It also sets out the commencement date and provides space for signature of the agreement.
	The frequency of maintenance visits should be entered in the first line. This should be selected to take into account the life of all consumables, so that they can be replaced during maintenance visits before they run out or deteriorate.
	The date of the first maintenance visit *should be specified by entering the maximum number of months in the second line.*
	The word "standard" or "premium" should be deleted or crossed through to indicate the chosen level of service.
	The first table "Maintenance charges" should be filled in with the agreed annual charge for the regular maintenance visits.
	The second table "Rates for additional services" should be filled in with the agreed charges for an emergency visit and an additional maintenance visit. These costs exclude VAT, which must be paid by the customer.
	The "emergency visit labour charge" is to cover the cost of an additional inspection requested by the customer to take place outside the defined working hours. The "additional maintenance visit charge" is to cover the cost of any additional maintenance visits requested by the customer to take place within the defined working hours. For these additional services, the same maintenance services will be required as for a standard maintenance visit.
	The commencement date is the date from which the agreement comes into force. *This should be filled in using the space provided. A representative of the customer and of the maintainer should sign and insert their name and date of signature in the spaces provided.*

Section 3 – Definitions	
	These clauses set out the definitions of the key terms used in the agreement.
	It should be noted that:
	• the commencement date is the date from which the agreement comes into force. *This should be filled in using the space provided in Clause 3.3.* This may or may not be the date of the agreement
	• consumables are the parts that the maintainer will be responsible for replacing during each maintenance visit
	• replacement parts are defined as all elements of the equipment that are not consumables.

Section 4 – Maintenance services	
Clauses 4.1 and 4.2	Section 4 sets out actions that the maintainer will be required to carry out on each maintenance visit. The specific actions required to maintain the equipment are set out in Schedule 1.
Clauses 4.3 and 4.4	Clause 4.3 sets out the requirement for a full inspection and replenishment of the consumables listed in Tables 2 and 3 of Schedule 1. Clause 4.4 sets out the responsibility of the maintainer for provision of the consumables listed in Table 2 of Schedule 1. The customer will be responsible for providing the other consumables free of charge.
Clauses 4.5, 4.6 and 4.7	The working hours in which the maintainer will carry out the visits are defined in Clause 4.5. *Details of the working hours should be entered in this clause.* Clause 4.6 sets out the timescale in which the maintainer must respond to requests for additional services. Clause 4.7 states that this agreement will take precedence over the maintainer's standard conditions of sale.
Section 5 – Exclusion from services	
	The actions that are excluded from the agreement can be summarised as: ● modifications to the equipment ● defects caused by misuse, neglect or failure to follow the instructions, except those caused by the maintainer.
Section 6 – Standard service OR premium service	
	This section describes the two alternative levels of service. *The section not relating to the chosen level of service should be crossed through or deleted.*
Standard service	The main conditions applying to the <u>standard</u> level of service are the following: ● the customer must pay for all additional maintenance visits and replacement parts (including labour) unless caused by the maintainer's negligence ● the maintainer must detail any replacement parts required in the maintenance report ● during additional maintenance visits the maintainer should identify and correct any faults. The cost of any replacement parts should be agreed with the customer prior to installation.
Premium service	The main conditions applying to the <u>premium</u> level of service are the following: ● the maintainer must supply, install and test all replacement parts within seven days of identifying the fault ● the maintainer will carry out additional maintenance visits if the customer reports a fault with the equipment ● the customer must pay for all emergency visits ● during additional maintenance visits the maintainer should identify and correct any faults.
Section 7 – Maintenance charges	
	These clauses set out the terms that apply to the payment of the maintenance charges.
Clause 7.1	The agreed charges will be paid every year in advance, on or before the commencement date, and on the same date of every following year that the agreement is kept in force. If the customer does not pay on time the maintainer may terminate the agreement as set out in Section 15.

Clause 7.2	The maintainer has the right to amend the charges for the following year by written notice given at least 30 days before the annual renewal of the agreement. The customer is under no obligation to renew the agreement for a further year.
	The maintainer may terminate the agreement if the customer does not pay the maintenance charges on time. The customer has 30 days from the due date to pay the maintenance charges.
Clause 7.3	The customer will be responsible for paying VAT on the charges and will not make any deductions from them.

Section 8 – Customer's obligations

	The customer's obligations under the agreement are to: • pay the charges promptly • operate the equipment properly • refrain from moving or modifying the equipment without consent • make any operation and maintenance records available to the maintainer • provide access • maintain associated drainage and pipework.

Section 9 – Maintainer's obligations

	The maintainer's obligations under the agreement are to: • carry out the maintenance in a proper, diligent and workmanlike manner • use appropriate equipment and competent staff • indemnify the customer against any losses due to the maintainer's neglect or default • rectify any breach of the agreement identified by the customer.

Section 10 – VAT

	VAT should be paid in addition to any charges unless otherwise stated.

Section 11 – Liability

	The maintainer's liabilities in connection with the failure of the equipment are limited to losses or damage caused by the maintainer's negligence.

Section 12 – Commencement and term of the agreement

	The agreement will run for a year at a time or until one party gives 30 days' prior written notice.
	If the customer moves house it is advised that they terminate the contract and notify the incoming owner/tenant of the agreement in advance to enable them to enter into a similar agreement.

Section 13 – Termination for breach

	The reasons for which the agreement may be terminated immediately can be summarised as: • the customer or maintainer fails to comply in all respects with the agreement • either party dies or becomes bankrupt • either party is subject to liquidation or receivership.

Section 14 – Termination consequences

	This section sets out the consequences of termination of the agreement.
	Note that if the agreement is terminated, the system should still be properly maintained. Continued use of the system when it is not being properly maintained will increase risks to health. If this is the case, the system should be decommissioned.

Clauses 14.1 and 14.2	Within 30 days of termination the maintainer must produce a final account taking into account: • any refund due to the customer for maintenance visits paid for in advance but not yet carried out. This should be calculated in proportion to the total number of visits paid for • all arrears due to the maintainer under this agreement. This account should be settled within 30 days of receipt of the final account.
Clauses 14.3 and 14.4	These clauses give both parties the entitlement to use the rights granted by the agreement, including enforcing the other party's liabilities or other common law rights available for redress as a consequence of breach of this agreement.

Section 15 – Sub-contracting

Clause 15.1	This clause sets out that the maintainer may sub-contract its obligations subject to the customer's prior written consent.
Clause 15.2	This clause sets out that the customer may only assign or delegate any of the rights of the agreement to another party with the prior written consent of the maintainer.

Section 16 – Third-party rights

	No other parties may acquire any rights from this agreement.

Section 17 – Discretion

	Any discretion or opinion exercised will be binding only if it is agreed in writing by both parties.

Section 18 – Variation

	No variation to the agreement will be valid unless it is agreed in writing by both parties.

Section 19 – Law and jurisdiction

	English law and the jurisdiction of English courts apply to this agreement.

Section 20 – Change of address

	Both parties must give notice of a change of address or contact detail at the earliest possible opportunity, within a maximum of 48 hours.

Section 21 – Notices

	These clauses set out the conditions that must be met for a notice to be considered to have been served.

Section 22 – Force majeure

	Either party whose actions are prevented by force majeure must give prompt notice and use their best endeavours to carry out the action, but will be excused if these fail.

Section 23 – Arbitration

	Parties may refer disputes to a commonly agreed independent arbitrator or, where one cannot be agreed upon, to an arbitrator nominated by the president of the Chartered Institution of Arbitrators.

COMMENTARY ON SCHEDULE I

The following section sets out guidance to Schedule 1 of the model agreement for rainwater and greywater use systems (RW/GW MA).

Section I – Details of the equipment	
Clause I	Details of the rainwater or greywater equipment should be included in this section.
	Any drawings, maintenance manuals or other guidance should be included as Appendix I and a brief description of this information provided.

Section 2 – Health, safety and environment	
Clauses 2.1 and 2.2	Maintenance of this equipment may be hazardous if not correctly managed. Risk assessments should be carried out by the maintainer in advance of carrying out this maintenance work, and all procedures should be strictly followed.
	A list of any site-specific precautions should be included in the Schedule. This should include any information that is particularly relevant to the site, such as the storage of hazardous substances. There are unlikely to be any site-specific precautions in the domestic environment.
	More information about risk assessments and health and safety in the workplace can be obtained from the Health and Safety Executive (see <www.hse.gov.uk>).
Clauses 2.3 and 2.4	These clauses relate to the environmental impact associated with the disposal of any consumables or replacement parts. These items should be disposed of safely and properly.
	Information about waste disposal is available from the Environment Agency website <www.environment-agency.gov.uk>.

Section 3 – Consumables	
Clauses 3.1 and 3.2	All consumables (eg filters, disinfectant) should be replaced at the frequency recommended by the manufacturer. The manufacturer's instructions should set this out in the maintenance information.
	Table 2 should be completed showing all consumables that are to be provided and replaced by the maintainer during the maintenance visits. The maintainer should charge the customer for the cost of the materials as part of the maintenance charge.
	Table 3 should be completed to show any consumables that are to be replaced by the maintainer during the maintenance visits but will be provided by the customer. It is possible that a large organisation may purchase the consumables directly from the manufacturer for replacement by the maintainer. In this case, the maintainer should charge only for the labour cost involved in making the replacement.

Section 4 – Water quality	
Clause 4.1 Rainwater use systems	This clause sets out the standards to which the rainwater should be treated, the tests required and the frequency of testing. *Table 4 should be completed with reference to Chapter 5 above, the manufacturer's instructions or other statutory guidance if it becomes available.*
	If the tests shown in Table 4 are not required, they should be removed.
	Standards for residual disinfectant are available only for chlorine. If another disinfectant is used, such as bromine, the residual disinfectant required should be in accordance with manufacturer's instructions.
Clause 4.1 Greywater	This clause sets out the standards to which the greywater should be treated, the tests that are required and the frequency of testing. *Table 4 should be completed with reference to Chapter 5 above or other statutory guidance if it becomes available.*
	If the tests shown in Table 4 are not required, they should be removed.
	Standards for residual disinfectant are available only for chlorine. If another disinfectant is used, such as bromine, the residual disinfectant required should be in accordance with manufacturer's instructions.

Clause 4.2	This clause sets out the requirements for carrying out the tests specified in Table 4.
Clause 4.3	It is important to keep a record of maintenance visits and all work carried out. This clause requires the maintainer to provide a maintenance report following each visit. *The time interval between the maintenance visit and the provision of the maintenance report should be included.*
Clause 4.4	This clause sets out the procedure to be followed should any of the water quality tests not conform to the standards set out in Table 4. The procedure is as follows. 1　The maintainer takes and analyses a further set of samples. 2　If these fail, the maintainer carries out remedial work to ensure the equipment is working correctly. 3　Further samples are tested to see if the equipment conforms. All testing and labour is at the expense of the maintainer.

Section 5 – Mechanical and electrical equipment	
Clauses 5.1 and 5.2	Certain tasks should be carried out during each visit to ensure that the mechanical equipment continues to function correctly. Where appropriate, all mechanical items shall be oiled, greased, resealed and left in a satisfactory condition by the maintainer. *The specific tasks maintenance tasks required should be included as Clause 5.2. It is likely that the* manufacturer will provide a list of maintenance tasks with the instructions.
Clauses 5.3 and 5.4	Certain tasks should be carried out during each visit to ensure that the electrical equipment continues to function correctly. It is essential that all items of electrical equipment are made safe following any maintenance work that is carried out. *The specific maintenance tasks required should be included as Clause 5.4. It is likely that the* manufacturer will provide a list of maintenance tasks with the instructions.

Section 6 – Water supplied to the equipment	
Clause 6.1	It is essential that no foul water sources be connected to the greywater or rainwater equipment.

Section 7 – Labelling of pipework	
Clauses 7.1 and 7.2	All pipework intended for reclaimed water should be labelled correctly in accordance with current WRAS guidance. This is fully described in Chapter 5 of the guidance. More information is available at <www.wras.co.uk>. Any changes that the maintainer makes to the greywater pipework shall be in accordance with the WRAS guidelines cited above. A clear diagram indicating the changes to the pipework shall be provided to the customer with the maintenance report.

Section 8 – Mains water back-up	
Clauses 8.1 and 8.2	If there is a meter fitted to the potable water back-up this should be read during each maintenance visit. High readings will indicate low use of the reclaimed water.

References and further information

BREWER, D, BROWN, R and STANFIELD, G (2001). *Rainwater and greywater in buildings – project report and case studies*. TN 7/2001, BSRIA, Bracknell

BROWN, R and PALMER, A (2002a). *Water reclamation guidance – design and construction of systems using grey water*. TN 6/2002, BSRIA, Bracknell

BROWN, R and PALMER, A (2002b). *Water reclamation guidance – laboratory testing of systems using grey water*. TN 7/2002, BSRIA, Bracknell

BS 1710:1984 *Specification for identification of pipelines and services*. British Standards Institution, London

DEFRA (2002). *Directing the flow*. Department for Environment, Food and Rural Affairs, London

DETR (1999). *A better quality of life, a strategy for sustainable development in the UK*. Department of the Environment, Transport and the Regions, London

DIXON, A, BUTLER, D and FEWKES, A (1999). "Guidelines for greywater reuse – health issues". *J CIWEM*, vol 13, no 5, pp 322–326

DTLR (2001a). *Development and flood risk*. PPG 25, Department for Transport, Local Government and the Regions, London

DTLR (2001b). *Building Regulations 2000. Approved Document H: drainage and waste disposal*. Stationery Office, London

DTLR (2002). *Part H of the Building Regulations*. Department for Transport, Local Government and the Regions, London

EA (1999). *A study of domestic greywater recycling*. Environment Agency, Bristol

EA (2001). *Conserving water in buildings*. Water Efficiency Fact Cards, Environment Agency, Bristol

EA (2003). *Harvesting rainwater for domestic uses: an information guide*. Environment Agency, Bristol

HEALTH AND SAFETY COMMISSION (2000). *Legionnaires disease. The control of legionella bacteria in water systems. Approved code of practice*. L8, HSE Books, Sudbury

LEGGETT, D, BROWN, R, BREWER, D, STANFIELD, G and HOLLIDAY, E (2001a). *Rainwater and greywater use in buildings. Best practice guidance*. C539, CIRIA, London

LEGGETT, D, BROWN, R, BREWER, D, STANFIELD, G and HOLLIDAY, E (2001b). *Rainwater and greywater use in buildings – decision-making for water conservation*. PR80, CIRIA, London

MARTIN, P, TURNER, B, WADDINGTON, K, PRATT, C, CAMPBELL, N, PAYNE, J and REED, B (2000a). *Sustainable urban drainage systems – design manual for Scotland and Northern Ireland*. C521, CIRIA, London

MARTIN, P, TURNER, B, WADDINGTON, K, DELL, J, PRATT, C, CAMPBELL, N, PAYNE, J and REED, B (2000b). *Sustainable urban drainage systems – design manual for England and Wales*. C522, CIRIA, London

MARTIN, P, TURNER, B, DELL, J, PAYNE, J, ELLIOTT, C and REED, B (2001). *Sustainable urban drainage systems – best practice manual*. C523, CIRIA, London

MARTIN, P, TURNER, B, WADDINGTON, K, DELL, J, PRATT, C, CAMPBELL, N, PAYNE, J, REED, B and ELLIOTT, C (2004). *SUDS compilation*. C599CD, CIRIA, London

PRATT, C WILSON, S and COOPER, P (2002). *Source control using constructed pervious surfaces. Hydraulic, structural and water quality performance issues*. C582, CIRIA, London

SAMUELS, P, WOODS, B, HUTCHINGS, C, FELGATE, J and MOBS, P (2004). *Sustainable water management in land use planning*. C630, CIRIA, London

USEPA (1992). *Guidelines for water reuse*. EPA/625/R-92/004, United States Environmental Protection Agency, Center for Environmental Research Information, Cincinatti

WILSON, S, BRAY, R and COOPER, P (2004). *Sustainable drainage systems – hydraulic, structural and water quality advice*. C609, CIRIA, London

WRAS (1999a). *Reclaimed water systems: information about installing, modifying or maintaining reclaimed water systems*. Information and guidance note no 9-02-04, Issue 1, Water Regulations Advisory Scheme, Oakdale, Gwent

WRAS (1999b). *Marking and identification of pipework for reclaimed (greywater) systems*. Information and guidance note no 9-02-05, Issue 1, Water Regulations Advisory Scheme, Oakdale, Gwent

LEGISLATION

Bathing Water Directive 1975 (76/160/EEC)

The Building Regulations 2000 (as amended) (SI 2000/2531)

The Confined Spaces Regulations 1997 (SI 1997/1713)

The Construction (Design and Management) (Amendment) Regulations 2000 (SI 2000/2380, amending SI 1994/3140)

The Control of Substances Hazardous to Health Regulations 1999 (SI 1999/437)

The Electricity at Work Regulations 1989 (SI 1989/635)

Health and Safety at Work, etc Act 1974

Local Government Act 2000 (2000 c. 22)

The Manual Handling Operations Regulations 1992 (SI 1992/2793)

The Personal Protective Equipment at Work Regulations 1992 (SI 1992/2966)

The Private Water Supplies Regulations 1991 (SI 1991/2790)

The Provision and Use of Work Equipment Regulations 1998 (SI 1998/2306)

Town and Country Planning Act 1990 (1990 c. 8)

Water Industry Act 1999 (1999 c. 9)

The Water Supply (Water Fittings) Regulations 1999 (SI 1999/1148)

The Workplace (Health, Safety and Welfare) Regulations 1992 (SI 1992/3004)

USEFUL WEBSITES

CIRIA	<www.ciria.org>
	<www.ciria.org/suds>
Defra	<www.defra.gov.uk/>
Environment Agency	<www.environment-agency.gov.uk/savewater>
Health and Safety Executive	<www.hse.gov.uk>
Office of the Deputy Prime Minister	<www.odpm.gov.uk>
Office of Water Services (Ofwat)	<www.ofwat.gov.uk>
Water Regulations Advisory Scheme	<www.wras.co.uk>